**Information Circular 9478**

# Significant Dust Dispersion Models for Mining Operations

By W. R. Reed, Ph.D., P.E.

DEPARTMENT OF HEALTH AND HUMAN SERVICES
Centers for Disease Control and Prevention
National Institute for Occupational Safety and Health
Pittsburgh Research Laboratory
Pittsburgh, PA

September 2005

## ORDERING INFORMATION

Copies of National Institute for Occupational Safety and Health (NIOSH)
documents and information
about occupational safety and health are available from

NIOSH–Publications Dissemination
4676 Columbia Parkway
Cincinnati, OH 45226–1998

FAX: 513–533–8573
Telephone: 1–800–35–NIOSH
(1–800–356–4674)
E-mail: pubstaft@cdc.gov
Web site: www.cdc.gov/niosh

*This document is the public domain and may be freely copied or reprinted.*

Disclaimer: Mention of any company or product does not constitute endorsement by NIOSH.

DHHS (NIOSH) Publication No. 2005–138

# CONTENTS

Page

| | |
|---|---|
| Abstract | 1 |
| Introduction | 2 |
| Health effects of dust | 2 |
| Regulations pertaining to dust | 3 |
|     Health and safety regulations | 3 |
|     Environmental regulations | 3 |
| Dust propagation models and mathematical algorithms | 4 |
|     Box model algorithm | 4 |
|     Gaussian model algorithm | 4 |
|     Eulerian model algorithm | 5 |
|     Lagrangian model algorithm | 5 |
|     Summary | 6 |
| Underground mine models | 6 |
|     Modeling with basic mathematical equations | 6 |
|         Hwang et al. study | 6 |
|         Courtney, Kost, and Colinet study | 7 |
|         Courtney, Cheng, and Divers study | 8 |
|         Bhaskar and Ramani study | 8 |
|         Xu and Bhaskar study | 9 |
|         Chiang and Peng study | 9 |
|     Modeling with computational fluid dynamics | 10 |
|         Wala et al. studies | 10 |
|         Brunner et al. study | 11 |
|         Heerden and Sullivan study | 11 |
|         Srinivasa et al. study | 11 |
| Surface mine models | 12 |
|     Mine-specific numerical models | 12 |
|         Cole and Fabrick study | 12 |
|         TRC environmental study | 13 |
|         Modeling of blasting phase studies | 13 |
|         Pereira et al. study | 13 |
|     ISC3 model | 13 |
|         ISC3 modeling technique | 14 |
|         EPA study | 14 |
|         Cole and Zapert study | 15 |
|         Reed et al. studies | 15 |
| Additional models | 16 |
| Summary | 17 |
| References | 18 |
| Appendix A.—Supporting equations used with the ISC3 model | 20 |

# ILLUSTRATIONS

| | |
|---|---|
| 1. Emissions from haul road as handled by the ISC3 model | 16 |
| 2. Representation of actual emissions from a moving haul truck | 16 |

# TABLES

| | |
|---|---|
| 1. Summary of dust dispersion models | 17 |

## UNIT OF MEASURE ABBREVIATIONS USED IN THIS REPORT

| | | | |
|---|---|---|---|
| ft | feet | mg | milligrams |
| $ft^2$ | square feet | $mg/ft^3$ | milligrams per cubic feet |
| ft/min | feet per minute | $mg/ft^2$-min | milligrams per square foot per minute |
| g/sec | grams per second | $mg/ft^3$-min | milligrams per cubic feet per minute |
| hr | hours | mg/L | milligrams per liter |
| km | kilometers | $mg/m^3$ | milligrams per cubic meter |
| m | meters | mph | miles per hour |
| m/sec | meters per second | sec | seconds |
| $m/sec^2$ | meters per second squared | $\mu m$ | micrometers |
| $m^2/sec$ | square meters per second | $\mu g/m^3$ | micrograms per cubic meter |
| $m^3/sec$ | cubic meters per second | | |

# SIGNIFICANT DUST DISPERSION MODELS FOR MINING OPERATIONS

By W. R. Reed, Ph.D., P.E.[1]

## ABSTRACT

Dust dispersion modeling is a subject that has had a large amount of research activity. Much of the research has focused on large-scale global or regional dispersion models. Other models have been created for industry-specific purposes. Furthermore, some of the past research has focused on dust dispersion modeling in the mining industry. This report presents the various dust dispersion models that have been developed specifically for the mining industry. The report first gives a brief background of the regulatory environment that helped to promote development of such models. It then presents an overview of the mathematical concepts used in this dispersion modeling. Finally, each of the various models developed for the mining industry are described, along with their associated mathematical algorithms and any field validation conducted on the different models.

[1] Mining engineer, Pittsburgh Research Laboratory, National Institute for Occupational Safety and Health, Pittsburgh, PA.

## INTRODUCTION

Historically and into the present day, mining operations have generated substantial quantities of airborne respirable dust, which has led to the development of lung disease in mine workers. Respirable dust consists of the particle size fraction of 4.0 $\mu$m or less [Lippman 1995]. Coal worker's pneumoconiosis (CWP) and silicosis are lung diseases that have adversely impacted the health of thousands of mine workers. Depending on the severity of the lung disease, symptoms range from reduced breathing capacity to death. During 1990–1999, more than 15,000 deaths resulted from CWP and more than 2,400 deaths resulted from silicosis in the mining industry [NIOSH 2002].

The primary means of controlling dust generation is through the application of ventilating air and water sprays. Ventilating air dilutes generated dust and moves dust away from mine workers. Water is applied during the cutting, transport, and crushing of the mined material in an effort to suppress dust liberated by these processes. Although significant advances in dust control technology have been realized, improved mining practices and equipment have meanwhile led to record production levels. Higher production levels have in turn resulted in the generation of additional dust. Ongoing research is searching for new and improved controls that can be used to reduce respirable dust exposures of mine workers. One tool that can be used to investigate dust generation and dispersion is computer modeling.

Modeling or simulation is a process whereby a system is created to simulate a real-life situation. Computer modeling is generally the most inexpensive and versatile method for analyzing a real-life situation and has become prevalent for solving problems related to physical processes, especially in research and development [Carroll 1987]. These simulations generally involve modeling a physical process and analyzing it through the use of a personal computer. This analysis involves trial-and-error methods applied to the model and tested with the actual physical process to perfect the model. Once this process is completed, the computer model can be used to identify problematic areas, and efforts can focus on finding solutions to address these particular concerns. Computer modeling of dust dispersion from mine sources can allow for the identification of potential hazard areas surrounding the source from a health and safety standpoint. It can also allow for the evaluation of dust control techniques to determine modifications necessary to improve dust control.

Many computer models have been created for predicting pollutant dispersion. However, the number of computer models for predicting dust dispersion at mining operations is small in comparison. Many of the mine-specific dust dispersion models are relatively unknown and therefore unused in the mining industry.

This report presents the significant dust dispersion models that are pertinent to the mining industry in an effort to highlight past and current dust modeling exercises. This review of mine-specific dust dispersion models also demonstrates how these modeling activities have sequentially improved upon existing models. The models are separated on the basis of their capability for use in either underground mining or surface mining. The models described can be particularly useful in the research of dust dispersion in the mining industry.

## HEALTH EFFECTS OF DUST

If silica is a component of respirable dust, then the effects of exposure pose a very serious health concern. Crystalline silica in respirable dust causes more than 250 U.S. workers to die each year of silicosis [U.S. Department of Labor 1996]. There are three levels of silicosis: chronic silicosis, which generally occurs after 10 years of exposure; accelerated silicosis, which generally occurs within 5 to 10 years of exposure; and acute silicosis, which can occur within a few weeks to 5 years of high exposure to silica [U.S. Department of Labor 1996]. Silicosis has no cure and is generally fatal. Miners are susceptible to silicosis both when working underground and when working on the surface.

CWP, or black lung, is a chronic disease occurring in miners that typically develops over a long period of respirable dust exposure (+20 years) and is generally fatal [NIOSH 1995]. Black lung, which occurs mainly in miners who work underground, is caused when respirable coal dust is a major component in the air that is breathed. Other employees who work coal stockpiles are also susceptible to black lung.

In addition to the hazards of respirable dust to humans, many epidemiologic studies show that particulate matter less than 10 $\mu$m in diameter ($PM_{10}$) also causes harm to humans. It has been shown that a 50-$\mu$g/m$^3$ increase in the 24-hr average $PM_{10}$ concentration was statistically significant in increasing mortality rates by 2.5%–8.5% [EPA 1996]. In relation to hospitalization due to chronic obstructive pulmonary disease, $PM_{10}$ caused a statistically significant increase by 6%–25% with an increase of the 24-hr average $PM_{10}$ concentration by 50 $\mu$g/m$^3$ [EPA 1996]. Other studies have shown that children are affected by short-term $PM_{10}$ exposure and that increased chronic cough, chest illness, and bronchitis were associated with a 50-$\mu$g/m$^3$ increase in the 24-hr average $PM_{10}$ concentrations [EPA 1996]. Long-term effects from $PM_{10}$ would depend on the amount of exposure to $PM_{10}$ over the life of a person.

There are other adverse impacts from $PM_{10}$ exposure in addition to the health effects. It is known that even small particles in the air hinder visibility, for the small particles scatter and absorb light as it travels to the observer from a source. This action results in extraneous light from sources other than the observed object being detected by the observer, thus impairing visibility [EPA 1996]. Climate change may also occur from $PM_{10}$ exposure because the small particles in the atmosphere absorb and reflect the radiation from the sun, affecting the cloud physics in the atmosphere [EPA 1996]. $PM_{10}$ may also have an effect on materials such as paint, wood, metals, etc. The effects depend on the amount of $PM_{10}$ in the atmosphere, the deposition of the $PM_{10}$ on the material, and the elemental composition of the $PM_{10}$ [EPA 1996].

# REGULATIONS PERTAINING TO DUST

There are two legislative acts that regulate the air quality from mining operations: the Federal Coal Mine Health and Safety Act of 1969, which was amended by the Federal Mine Safety and Health Act of 1977, and the Clean Air Act of 1970, which was amended in 1977 and 1990. The Federal Coal Mine Health and Safety Act of 1969 established the amount of dust allowable in air for health and safety purposes. The Clean Air Amendment of 1990 regulates air emissions from facilities from an environmental perspective.

## HEALTH AND SAFETY REGULATIONS

The Federal Coal Mine Health and Safety Act of 1969 established a limit of 2.0 mg/m³ for respirable dust for underground and surface coal mining operations [NIOSH 1995]. If more than 5% quartz or silica is found in the respirable dust, then the limit is determined by using the following formula (30 CFR[2] 71.101):

$$\Phi = \frac{10}{\% \ quartz} \quad (1)$$

where   $\Phi$ = respirable dust limit (mg/m³)
        % quartz = percent quartz or silica found in dust as a fraction

The American Conference of Governmental Industrial Hygienists also recommends this limit for respirable dust. For metal/nonmetal mining, a respirable dust standard is not enforced unless the level of quartz exceeds 1%. In these cases, a respirable dust standard is calculated by using the following formula:

$$\Phi = \frac{10}{\% \ quartz + 2} \quad (2)$$

The 2.0 mg/m³ limit was based on studies conducted on miners in the United Kingdom. This limit was intended to prevent the formation of CWP over the working life of a miner. However, additional studies have found that the risk of CWP based on these regulations is higher than originally estimated, and the National Institute for Occupational Safety and Health has recommended lowering the limit from 2.0 mg/m³ to 1.0 mg/m³ [NIOSH 1995].

[2] Code of Federal Regulations. See CFR in references.

## ENVIRONMENTAL REGULATIONS

The Clean Air Amendment of 1990 regulates emissions from any facility into the air and addresses toxic substances. It also creates the national ambient air quality standards (NAAQS) for the criteria pollutants: carbon monoxide (CO), oxides of nitrogen ($NO_x$), sulfur oxides ($SO_x$), volatile organic compounds (VOC), lead (Pb), and $PM_{10}$ [Schnelle and Dey 2000]. NAAQS have been in effect for $PM_{10}$ since before 1987 [Watson et al. 1997]. As dictated by the standards, facilities are not allowed to emit levels of $PM_{10}$ pollutants at a rate that causes the following standards to be exceeded [Watson et al. 1997]:

- Twenty-four hour average $PM_{10}$ not to exceed 150 $\mu g/m^3$ for a 3-year average of annual 99th percentiles at any monitoring site in a monitoring area.

- Three-year average $PM_{10}$ not to exceed 50 $\mu g/m^3$ for three annual average concentrations at any monitoring site in a monitoring area.

The NAAQS regulations for $PM_{10}$ are the maximum emission levels allowable in ambient air; however, states have the right to create stricter regulations. For example, in California the 24-hr average for $PM_{10}$ is 50 $\mu g/m^3$ and its annual average is 30 $\mu g/m^3$ [California Air Resources Board 2003].

The NAAQS are also used to determine if an area can be classified as a nonattainment area. In determining nonattainment areas, each state is required to have in place an air monitoring network for different regions [Watson et al. 1997]. This air monitoring network measures the air for a particular pollutant. If the NAAQS cannot be met for one or more pollutants, then the region is designated as "nonattainment" for that pollutant. Nonattainment areas can have stricter standards applied for that region, and permittees may have to institute better pollution control technology at their facilities in order to obtain approvals for air quality permits [Virginia Department of Environmental Quality 1996]. Overall, the NAAQS regulations for $PM_{10}$ are much stricter than the 2.0 mg/m³ respirable dust limit imposed by the Mine Safety and Health Administration (MSHA) regulations. However, the NAAQS regulations pertain to ambient air, which is defined as air that is accessible to the general public. Therefore, the NAAQS would not apply within the boundaries of areas that are enclosed by a fence.

Modeling is generally required for two reasons:

(1) *During new construction or any significant modifications/expansions of an existing facility that results in an emissions*

*increase above the state or federal limits.* This threshold amount can vary from state to state. For example, in Virginia, if a facility emits more than 250 tons of $PM_{10}$ per year, then modeling of the emissions from the facility is required in order to obtain a permit [Virginia Department of Environmental Quality 1996]. Georgia requires that any facility that emits more than 100 tons of a pollutant per year become a Title V facility, which may require emissions modeling of the facility [Georgia Department of Natural Resources 1994]. Title V of the Clean Air Act pertains to regulations regarding emissions of pollutants from a facility. All operations, including mining and quarrying, that have the potential to emit greater than 250 tons per year of a regulated pollutant are required to obtain a Title V operating permit. Additionally, for 28 specific industrial categories, the threshold is only 100 tons per year. The individual states may have more stringent regulations and can require all facilities to be subjected to the 100-ton-per-year threshold and may require modeling of emissions [Heinerikson 2004].

(2) *An ambient air monitor records a violation of an NAAQS.* The state or other regulatory authority may conduct modeling to determine the cause of the violation and use the model to develop a control strategy to prevent future occurrences of NAAQS violations. Therefore, modeling may be an important part of obtaining an air quality permit, depending on the amount of $PM_{10}$ emitted by the facility.

## DUST PROPAGATION MODELS AND MATHEMATICAL ALGORITHMS

The results from modeling the emissions of a facility are used to ensure that the regional air quality does not exceed the NAAQS or deteriorate the air quality further [Schnelle and Dey 2000]. If the modeling results show the facility will not cause the regional air quality to exceed the NAAQS or deteriorate the air quality, then the air quality permit will be granted. Otherwise, the air quality permit application will be denied. Therefore, it is important that the modeling method accurately estimate both the amount of pollutant a facility will emit and the pollutant's dispersion. The use of a modeling method that overestimates the ambient concentrations resulting from emissions sources at the facility may result in denial of an air quality-related permit, just as an underestimation may result in an inappropriate permit issuance.

Modeling of pollutant dispersion is completed using mathematical algorithms. There are several basic mathematical algorithms in use: the box model, Gaussian model, Eulerian model, and Lagrangian model [Collett and Oduyemi 1997].

### BOX MODEL ALGORITHM

The box model is the simplest of the modeling algorithms. It assumes the airshed is in the shape of a box. The air inside the box is assumed to have a homogeneous concentration. The box model is represented using the following equation:

$$\frac{dCV}{dt} = QA + uC_{in}WH - uCWH \quad (3)$$

where  $Q$ = pollutant emission rate per unit area
$C$ = homogeneous species concentration within the airshed
$V$ = volume described by box
$C_{in}$ = species concentration entering the airshed
$A$ = horizontal area of the box ($L \times W$)
$L$ = length of the box
$W$ = width of the box
$u$ = wind speed normal to the box
$H$ = mixing height

Although useful, this model has limitations. It assumes the pollutant is homogeneous across the airshed, and it is used to estimate average pollutant concentrations over a very large area. This mathematical model is very limited in its ability to predict dispersion of the pollutant over an airshed because of its inability to use spatial information [Collett and Oduyemi 1997].

### GAUSSIAN MODEL ALGORITHM

The Gaussian models are the most common mathematical models used for air dispersion. They are based upon the assumption that the pollutant will disperse according to the normal statistical distribution. The Gaussian equation generally used for point-source emissions is given as follows:

$$\chi = \frac{Q}{2\pi u_s \sigma_y \sigma_z} \left[ \exp\left\{-0.5\left(\frac{y}{\sigma_y}\right)^2\right\} \right] \left[ \exp\left\{-0.5\left(\frac{H}{\sigma_z}\right)^2\right\} \right] \quad (4)$$

where  $\chi$ = hourly concentration at downwind distance $x$
$Q$ = pollutant emission rate
$u_s$ = mean wind speed at release height
$\sigma_y, \sigma_z$ = standard deviation of lateral and vertical concentration distribution
$y$ = crosswind distance from source to receptor
$H$ = stack height or emission source height

The terms $\sigma_y$ and $\sigma_z$ are the standard deviations of the horizontal and vertical Gaussian distributions that are used to represent the plume of the pollutant. These coefficients are based upon the atmospheric stability coefficients created by Pasquil and Gifford, and they generally become larger as the distance downwind from the source becomes greater. Larger standard deviations mean that the Gaussian curve or plume has a low peak and a wide spread; smaller standard deviations mean that the Gaussian curve or plume has a high peak and a narrow spread [Oduyemi 1994].

When using this equation for calculation of pollutant dispersion, there are some assumptions that must be made in order for the equation to be valid: (1) the emissions must be constant and uniform, (2) the wind direction and speed are constant, (3) downwind diffusion is negligible compared to vertical and crosswind diffusion, (4) the terrain is relatively flat, i.e., no crosswind barriers, (5) there is no deposition or absorption of the pollutant, (6) the vertical and crosswind diffusion of the pollutant follow a Gaussian distribution, (7) the shape of the plume can be represented by an expanding cone, and (8) the use of the horizontal and vertical standard deviations, $\sigma_y$ and $\sigma_z$, requires the turbulence of the plume to be homogeneous throughout the entire plume [Beychok 1994].

The accuracy of this model to predict pollutant concentrations has been documented to be within 20% for ground level emissions at distances less than 1 km. For elevated emissions, the accuracy is within 40%. At distances greater than 1 km, the equation is estimated to be accurate within a factor of 2. The Gaussian model also has the limitation that it cannot be used for subhourly prediction of concentrations [Collett and Oduyemi 1997].

## EULERIAN MODEL ALGORITHM

Eulerian models solve a conservation of mass equation for a given pollutant. The equation generally follows the form [Collett and Oduyemi 1997]:

$$\frac{\partial \langle c_i \rangle}{\partial t} = -\overline{U} \cdot \nabla \langle c_i \rangle - \nabla \cdot \langle c_i' U' \rangle + D \nabla^2 \langle c_i \rangle + \langle S_i \rangle \quad (5)$$

where  $U = \overline{U} + U'$
 $U$ = windfield vector $U(x,y,z)$
 $\overline{U}$ = average wind field vector
 $U'$ = fluctuating wind field vector
 $c = \langle c \rangle + c'$
 $c$ = pollutant concentration
 $\langle c \rangle$ = average pollutant concentration; $\langle \ \rangle$ denotes average
 $c'$ = fluctuating pollutant concentration
 $D$ = molecular diffusivity
 $S_i$ = source term

The wind field vector $U$, which is normally used, is considered turbulent and results in $\overline{U}$ and $U'$, which are the components of the turbulent wind field vector being used in Equation 5. The turbulent wind field vector also affects the pollutant concentration $c$ in a similar manner with the terms $\langle c \rangle$ and $c'$. The term representing molecular diffusivity is neglected as the magnitude of this term is significantly small. When the rate of diffusion is assumed to be constant, the turbulent diffusion term $\nabla \cdot \langle c_i' U' \rangle$ is modeled as $\langle c_i' U' \rangle = -K \nabla \langle c_i \rangle$, where $K$ is an eddy diffusivity tensor. This tensor is simplified so that diffusivity transport is along the turbulent eddy vector, making the eddy diffusivity tensor diagonal and the cross vector diffusivities negligible, i.e.,

$$K = \begin{bmatrix} K_{xx} & 0 & 0 \\ 0 & K_{yy} & 0 \\ 0 & 0 & K_{zz} \end{bmatrix}, \text{ where } K_{xx} = K_{yy} = K_H, \text{ with } K_H \text{ being}$$

horizontal diffusivity [Collett and Oduyemi 1997].

Equation 5 can be difficult to solve because the advection term $-\overline{U} \cdot \nabla \langle c_i \rangle$ is hyperbolic, the turbulent diffusion term is parabolic, and the source term is generally defined by a set of differential equations. This type of equation can be computationally expensive to solve and requires some form of optimization in order to reduce the solution time required. Solutions have been achieved by reducing the problem to one and two dimensions rather than using three dimensions [Collett and Oduyemi 1997].

## LAGRANGIAN MODEL ALGORITHM

Lagrangian models predict pollutant dispersion based on a shifting reference grid. This shifting reference grid is generally based on the prevailing wind direction, or vector, or the general direction of the dust plume movement. The Lagrangian model has the following form:

$$\langle c(r,t) \rangle = \int_{-\infty}^{t} \int p(r,t|r',t') S(r',t') \, dr' \, dt' \quad (6)$$

where  $\langle c(r,t) \rangle$ = average pollutant concentration at location $r$ at time $t$
 $S(r',t')$ = source emission term
 $p(r,t|r',t')$ = the probability function that an air parcel is moving from location $r'$ at time $t'$ (source) to location $r$ at time $t$

The probability function must be derived as a function of the prevailing meteorology, which is appropriate for sources consisting of gases. If the source of emissions consists of particles, then more information must be incorporated into the function, such as the particle size distribution and the particle density [Collett and Oduyemi 1997].

This mathematical model has limitations when its results are compared with actual measurements. This is due to the dynamic nature of the model. Measurements are generally made at stationary points, while the model predicts pollutant concentration based upon a moving reference grid. This makes it difficult to validate the model during initial use. To compensate for this problem, the Lagrangian models are typically modified by adding an Eulerian reference grid. This allows for better comparison to actual measurements because it incorporates a static reference grid into the model [Collett and Oduyemi 1997].

## SUMMARY

These four mathematical models are the basic approaches used for air dispersion modeling. There are many variations based upon these equations. Some variations add statistical functions to represent the randomness of wind direction, wind speed, and turbulence; others include the introduction of site-specific source terms. Because of the increased computational power available via personal computers, the models have become more complex and varied. This has resulted in the creation of a vast number of computer models for air dispersion.

## UNDERGROUND MINE MODELS

Air dispersion modeling has not bypassed the mining industry. A number of mine-specific models have been created. Most of the models that have been validated with testing at mine sites are those developed for underground mines. The focus of many of the underground mining studies validating the models was on respirable dust. However, these models also have the positive feature of having the ability to predict dust dispersion of all dust particle size fractions.

### MODELING WITH BASIC MATHEMATICAL EQUATIONS

#### Hwang et al. Study

Hwang et al. [1974] discussed several models for predicting dust dispersion in an underground entry by a turbulent gas stream, in other words, "the prediction of dust dispersion after an explosion." The basic diffusion equation was defined as:

$$\frac{\partial c}{\partial t} + U \frac{\partial c}{\partial z} = k \left( \frac{\partial^2 c}{\partial x^2} + \frac{\partial^2 c}{\partial y^2} + \frac{\partial^2 c}{\partial z^2} \right) \quad (7)$$

where  $c$ = dust concentration (mg/L)
$U$ = convection velocity (units unknown)
$k$ = diffusion coefficient (units unknown)
$x,y,z$ = directions of coordinate grid (units unknown)
$t$ = time (units unknown)

This equation can be described as using an Eulerian approach, as it addresses the conservation of mass and assumes a stationary reference grid. From Equation 7, this report derived mathematical interpretations of the modeling process for four different types of sources: point source, line source, moving line source, and flat plane. These resulting derivations are Gaussian in nature, suggesting that the authors shift from an Eulerian to a Gaussian approach. The resulting modeling equations for each type of source are rather lengthy and thus not fully explained here. For example, the equation for an instantaneous point source in the plane $z = z_1$ at the point $(x_1, y_1)$ emitted at time $t = t_1$ is given below (Equation 8):

$$(8) \quad c = \frac{Q e^{\frac{-\{(z-z_1)-U(t-t_1)\}}{4k(t-t_1)}}}{2ab\{\pi k(t-t_1)\}^{\frac{1}{2}}} \left\{ 1 + 2\sum_{m=1}^{\infty} e^{\frac{-km^2\pi^2(t-t_1)}{a^2}} \cdot \cos\frac{m\pi x}{a} \cos\frac{m\pi x_1}{a} \right\} \cdot \left\{ 1 + 2\sum_{n=1}^{\infty} e^{\frac{-kn^2\pi^2(t-t_1)}{b^2}} \cdot \cos\frac{n\pi y}{a} \cos\frac{n\pi y_1}{a} \right\}$$

where  c = dust concentration (mg/L)
U = convection velocity (units unknown)
k = diffusion coefficient (units unknown)
x,y,z = directions of coordinate grid (units unknown)
t = time (units unknown)
a = entry opening height (units unknown)
b = entry opening width (units unknown)
n = distance in a direction normal to the boundary or walls of the opening (assumed to be for the y direction) (units unknown)
m = undefined, but assumed to be similar to n except for the x direction (units unknown)
Q = point-source emission strength (units unknown)

Variables n and m are not very well defined, as the explanations for these variables were insufficient to fully describe what was meant by these variables. Results of calculations of these equations were completed, but no comparisons of calculated results to actual results were given as it was stated that there were no measured observations available [Hwang et al. 1974].

### Courtney, Kost, and Colinet Study

Courtney, Kost, and Colinet completed a study in 1982 that defined dust deposition in underground coal mine airways [Courtney et al. 1982]. The main emphasis of this study was to determine an optimum schedule for rock dusting entries in an underground coal mine by using an airborne particle deposition model. Testing was completed at eight locations in five U.S. underground coal mines. The deposition model in this study was based on a model created by Dawes and Slack in 1954. Their model was based on the deposition of coal dust in a small laboratory wind tunnel. The resulting model, which has a Gaussian character, is defined as:

$$\frac{\partial m}{\partial t} = Kc = Kc_0 \exp\left(\frac{-Kx}{vH}\right) \quad (9)$$

where  $\frac{\partial m}{\partial t}$ = dust deposition rate (units unknown)

x = distance of deposit from the dust source (units unknown)
$c_0, c$ = airborne dust concentration of particles of diameter D at the dust source and at x, respectively (units unknown)
v = air velocity (units unknown)
H = height of airway (units unknown)
K = rate constant, taken as Stoke's sedimentation velocity

$$K = kD^2 \quad (10)$$

where
D = particle diameter (units unknown)
k = Stoke's sedimentation constant (units unknown)

$$k = \frac{(\rho - \sigma)g}{18\eta} \quad (11)$$

where
$\rho$ = particle density (units unknown)
g = acceleration of gravity (units unknown)
$\sigma$ = density of air (units unknown)
$\eta$ = viscosity of air (units unknown)

This model was found to have satisfactory results for particles with diameters less than 40 $\mu$m, and the exponential decay with distance agreed with the experimental results [Courtney et al. 1982]. The study by Courtney, Kost, and Colinet also stated that Bradshaw and Godbert completed a study of the deposition rate of dust in the return airway of underground coal mines. The results of this study showed an exponential decay rate, but the first 23 m from the source were found to have two to four times more dust deposition than was calculated [Courtney et al. 1982]. In studies of dust deposition in underground coal mines, Ontin found that the deposition rate also decayed exponentially and that 50% of the airborne dust settled within 1.8 m of the source [Courtney et al. 1982]. Experimental testing demonstrated that Equation 9 may underpredict the deposition rate of dust at distances close to sources. Through testing, Courtney, Kost, and Colinet found that the deposition rate in pounds per square foot per hour was independent of the airborne particle size, but increased with increasing total airborne dust concentration. Their recommended deposition model, which is also Gaussian based, was as follows:

$$\frac{\partial m}{\partial t} = \left[\left(\frac{K_1 V}{S}\right)c_0\right]\exp\{-Ax\} \quad (12)$$

where  $\frac{\partial m}{\partial t}$ = dust deposition rate (units unknown)

$K_1$ = a proportionality constant, found to be 15.6 in this study (units unknown)
A = $K_1/v$
x = distance along the airway (units unknown)
$c_0$ = initial dust concentration (units unknown)
v = air velocity (units unknown)
V/S = volume/surface area of the airway (units unknown)

This equation was stated to be correct if the airflow is turbulent in the airway and not laminar and if the rate of deposition is exponential with distance [Courtney et al. 1982]. The study found that this model could be used for determining an optimum rock dusting schedule for an underground coal mine, but that further testing should be completed at many other mine sites because of the variability from one mine location to another [Courtney et al. 1982].

## Courtney, Cheng, and Divers Study

Courtney, Cheng, and Divers proposed another deposition model for underground coal mines in a later study [Courtney et al. 1986]. The study stated that the "rate of decrease of the airborne concentration must be equal to the deposition of the airborne particles onto the surfaces of the airway" [Courtney et al. 1986]. The model was developed using an Eulerian approach, since this statement is analogous to the conservation of mass and the model uses a stationary reference grid in underground entries. The model is represented by the following equation:

$$-vA\frac{\partial c}{\partial x} = L\frac{\partial m}{\partial t} \qquad (13)$$

where  $v$ = air velocity (ft/min)
$A$ = cross-sectional area of airway (ft$^2$)
$c$ = local dust concentration (mg/ft$^3$)
$x$ = distance along airway (ft)
$\frac{\partial m}{\partial t}$ = rate of dust deposition per unit area along airway (mg/ft$^2$-min)
$L$ = deposition surface across airway (ft)
$L$ = perimeter if dust deposits on roof, walls, and floor
$L$ = width of airway if dust deposits only on floor

If the rate of dust deposition was dependent upon local dust concentration as stated in the study by Courtney, Kost, and Colinet, then Equation 13 could be represented as:

$$-Av\frac{\partial c}{\partial x} = Lkc \qquad (14)$$

where the terms are the same as given in Equation 13
$k$ = dust deposition rate constant (ft/min)

$$\frac{c}{c_0} = \exp\left\{\left(\frac{-L}{AV}\right)kx\right\} \qquad (15)$$

where  $c_0$ = dust concentration at the source (mg/ft$^3$)
$c$ = dust concentration at a distance $x$ from the source (mg/ft$^3$)

Equation 14 continues to use the Eulerian approach, but Equation 15 introduces a Gaussian component to the model.

Experiments to test the deposition of dust with varying air velocities and relative humidities were conducted at an underground limestone mine. It was thought that deposition might depend on Stokes' sedimentation velocity. However, it was found that the deposition was dependent upon air velocity and that large and small particles deposited at similar rates along the first 300 ft away from the source in the airway. The larger particles had fully deposited by 500 ft away from the source. The rough surface of the walls of the limestone mine were thought to affect the deposition of smaller particles by trapping the larger particles. The dependence of particle deposition on air velocity in the airway implied a change in the airborne particle size distribution, which remained to be explained [Courtney et al. 1986].

The results of the study demonstrated that the median particle sizes were higher at the floor of the airway (6.5 $\mu$m) than at the roof of the airway (4.7 $\mu$m) at 100 ft away from the source. At distances of 500–700 ft away from the source, the median diameters were closer together (4.5–4.9 $\mu$m). Respirable dust deposition rate was shown to decrease as a function of distance from the source. At low air velocities, the deposition rates were linear. At higher air velocities, the deposition rates decreased as the distance from the source became greater. Relative humidity was found to have a negligible effect on the dust deposition rate [Courtney et al. 1986].

Ratios of deposition rates of dust onto the floor, walls, and roof of the airways were also presented. These deposition rates were dependent upon particle size, and the floor deposition rate was greater than the roof and wall deposition rate. The ratios were established by studies done by Pereles and Owen [Courtney et al. 1986].

## Bhaskar and Ramani Study

Bhaskar and Ramani [1989] described a modeling method for the deposition of respirable dust in an underground coal mine. Their report is related to Bhaskar's Ph.D. dissertation entitled "Spatial and Temporal Behavior of Dust in Mines: Theoretical and Experimental Studies," completed at The Pennsylvania State University in 1987 [Bhaskar 1987]. Using the Eulerian method, the mathematical model presented was defined as:

$$\frac{\partial c}{\partial t} = E_x\left(\frac{\partial^2 c}{\partial x^2}\right) - U\frac{\partial c}{\partial x} + \text{sources} - \text{sinks} \qquad (16)$$

where  $c$ = concentration of airborne dust (units unknown)
$t$ = time (units unknown)
$E_x$ = dispersion coefficient (units unknown)
$x$ = distance from source (units unknown)
$U$ = velocity of airflow (units unknown)

The "source" term represents dust generated by cutting mechanisms in the underground mine; the "sink" term refers to the deposition of the dust on the floor, walls, and roof of the airway [Bhaskar and Ramani 1989]. This mathematical model is applied to all the particle size intervals that are represented in a dust cloud generated from a mining operation.

Results of comparing the model to experiments conducted in an underground airway showed that the model predicted deposition of the dust in airways satisfactorily. The model tended to predict better at lower airway air velocities than higher velocities. Also, total dust size was better predicted than respirable dust size [Bhaskar 1987].

Detailed explanations for the processes used in creating this model are given by Ramani and Bhaskar [1988]. The processes considered are particle deposition, deposition by convective diffusion, deposition due to gravity, coagulation, collision mechanisms, and reentrainment [Ramani and Bhaskar 1988]. Particle deposition is related to mass transfer of a particle to the immediate adjacent surface; this represents deposition onto the roof and walls of the airway and is represented by Brownian diffusion, eddy diffusion, or sedimentation. Deposition by convective diffusion refers to deposition caused by eddies in turbulent flow and represents deposition onto the walls. Deposition due to gravity uses the particle's gravitational velocity to determine the deposition of the particle onto the floor of the airway. Coagulation and collision mechanisms are related and are based upon the interaction of the particles with one another. These two processes are important in determining the airborne particle size distribution and therefore important in determining the amount of dust deposited onto the airway surfaces. They take into account forces such as electrostatic charge, Van der Waals forces, and the nature of the colliding particle's surfaces. Reentrainment evaluates the amount of dust that is generated from dust that has already been deposited. Dust may be reentrained due to the shear forces from the velocity of air in the airway exceeding the cohesive force of the particle on the surface. This process is dependent upon the air velocity in the airway [Ramani and Bhaskar 1988].

### Xu and Bhaskar Study

Xu and Bhaskar [1995] built on the work of Ramani and Bhaskar and conducted additional studies to further characterize particle deposition in underground coal mine entries. Better understanding of this process would allow for the formation of improved modeling techniques in the future. These studies determined the deposition velocities for coal dust in an underground mine airway. Deposition velocity consists of two components in turbulent flow: deposition velocity due to gravity and deposition velocity due to diffusion or turbulent deposition. Turbulent deposition occurs from the breakaway of eddies from the main airflow through the airway. Particles are transported to the wall by these eddies and have sufficient momentum to penetrate the stagnant air along the wall and remain at this location. Turbulent deposition, which is stated to be dominant for fine particles, was shown to be independent of particle size and airflow, but dependent upon particle density as air velocity increased. In contrast, the particle properties and air velocities influence gravitational deposition velocities more than they affect turbulent deposition velocities [Xu and Bhaskar 1995].

### Chiang and Peng Study

Chiang and Peng [1990] worked on the development of dust distribution models for underground coal longwall faces. Their approach used a statistical technique, which includes a Gaussian component, to predict dust concentrations at different locations along a longwall face. Their resulting report [Chiang and Peng 1990] describes the forces that have an impact on a dust particle and is the first mining research report to include deposition velocities in stationary air. The research also included suspended time and traveling distances of these same particle sizes when air velocity is 2.5 m/sec in an entry having a height of 2 m.

Chiang and Peng's report describes field studies conducted at several underground coal longwall mines to determine airflow patterns both in the plan and cross-section views along the longwall face. As within pipeflow, the maximum airflow velocity is generally located in the center of the longwall crosscut, but can be shifted to one side or the other depending on the location of other equipment (shearer, supports, and conveyor) in the crosscut [Chiang and Peng 1990].

A model of the airflow distribution was created, resulting in the following polynomial equation [Chiang and Peng 1990]:

$$f_x(Y,Z) = A + BY + CZ + DY^2 + EZ^2 + FYZ \qquad (17)$$

where  $Y$ = horizontal axis along the floor originating at the face and directed toward the gob (m)
$Z$ = vertical axis along the coal face and directed toward the roof (m)

$A$, $B$, $C$, $D$, $E$, and $F$ are coefficients that represent proportionality constants that were derived from regression analysis of measured velocity readings from different locations in and along the longwall crosscut.

The airflow model led to the development of the longwall dust dispersion model. This dispersion model uses the following equation to calculate the time-weighted-average dust concentration in the entryway [Chiang and Peng 1990]:

$$\overline{N}(x) = \int_0^{20} \overline{N}(D,x)\, dt \qquad (18)$$

where

$$N(D,x) = \frac{1}{T_c} \left[ \frac{1 - e^{-MD^2 x}}{MD^2} \left( \left(\frac{N_{01}(D)}{V_1}\right) + \left(\frac{N_{02}(D)}{V_2}\right) \right) + N_c(D)(L-x)\left(\frac{1}{V_1} + \frac{1}{V_2}\right) + N_c(D) T_d \right] \qquad (19)$$

where  $\bar{N}$ = time-weighted-average concentration at location $x$ (mg/m³)
$T_c$ = mean time of a complete mining cycle (sec)
$D$ = particle size (μm)
$L$ = length that shearer must travel to complete one cut (m)
$V_1, V_2$ = velocity of the shear: $V_1$ = head to tail-tail entry; $V_2$ = tail to headgate entry (m/sec)
$T_d$ = mean idle time for a complete mining cycle (sec)
$N_{01}(D)$ = size composition function of the source at the cross-section right behind the shearer on the return side for the head-to-tailgate entry trip (μm)
$N_{02}(D)$ = size composition function of the source at the cross-section right behind the shearer on the return side for the tail-to-headgate entry trip (μm)
$N_c(D)$ = size composition function of the source at the cross-section on the intake side of the shearer (assumed to be constant) (μm)

$$M = \frac{Apgm}{V_a u} \quad (20)$$

where $p$ = specific weight of dust particles (mg)
$g$ = acceleration of gravity (m/sec²)
$A$ = ratio of entry perimeter to effective ventilation air (units unknown)
$V_a$ = mean air velocity (m/sec)
$u$ = air viscosity (units unknown)
$m$ = unknown (units unknown)

In order to use these equations to calculate the time-weighted-average dust concentration at a location, four assumptions were made [Chiang and Peng 1990]:

1. The dust generation sources in the longwall are considered to be generating dust at a constant rate.
2. The shearer cuts coal in the head-to-tailgate direction and cleans the bottom from the tail-to-headgate direction. The supports are always advanced during the clean trip, and the distance from the shearer and the advancing support is constant throughout the trip.
3. The dust concentration is higher on the return side than the intake side.
4. The velocity of the shearer is constant in either direction.

It should also be noted that the integration limits of 0 to 20 in Equation 18 represent the particle size classes from 0 to 20 μm. This model was tested with data from the field studies that were completed to create the airflow distribution model and agreed well with the field study results [Chiang and Peng 1990]. It was found that air velocity is very important in dust particle distribution. Increasing the entry airflow and the dust generation from the source by corresponding magnitudes also increased the dust concentrations in the entryway. However, increasing the dust generation from the source by an amount that is half the magnitude of the entry airflow increase lowered the dust concentrations in the entryway [Chiang and Peng 1990]. This showed that eliminating the dust generation was very important in reducing the dust concentrations in the entryways. Changes in the specific weights of coal were also reviewed with this model, and it was found that no significant changes in dust concentrations occurred [Chiang and Peng 1990].

## MODELING WITH COMPUTATIONAL FLUID DYNAMICS

### Wala et al. Studies

Wala et al. [1997, 2001] describe using computational fluid dynamics (CFD) to characterize airflow in mine entries. CFD is a numerical analysis method used to solve fluid flow problems with a computer. This method generally follows an Eulerian approach, which is applied to the airflow modeling [Anderson 1995]. However, it can also incorporate the Lagrangian algorithm. The Lagrangian method is applied to the particles that are subjected to forces of gravity and airflow. The forces (gravity and airflow) are based on an Eulerian reference frame, whereas the Lagrangian algorithm is used to characterize the advection and diffusion processes that occur among the individual particles and influence each particle's trajectory [Collett and Oduyemi 1997].

The basis of CFD is the fluid flow motion equations for a Newtonian fluid, shown in tensor form:

$$\frac{\partial \rho}{\partial t} + \frac{\partial (\rho u_i)}{\partial x_i} = 0 \quad (21)$$

$$\rho \frac{\partial u_j}{\partial t} + \rho u_k \frac{\partial u_j}{\partial x_k} = \left[ -\frac{\partial p}{\partial x_j} + \frac{\partial \left( \lambda \frac{\partial u_k}{\partial x_k} \right)}{\partial x_j} + \frac{\partial \left( \mu \left[ \frac{\partial u_i}{\partial x_j} + \frac{\partial u_j}{\partial x_i} \right] \right)}{\partial x_i} + \rho f \right] \quad (22)$$

$$\rho\frac{\partial e}{\partial t}+\rho u_k\frac{\partial e}{\partial x_k}=\begin{bmatrix}-p\dfrac{\partial u_k}{\partial x}+\dfrac{\partial\left(k\dfrac{\partial T}{\partial x_j}\right)}{\partial x_j}+\\ \lambda\left(\dfrac{\partial u_k}{\partial x_k}\right)^2+\mu\left(\dfrac{\partial u_i}{\partial x_j}+\dfrac{\partial u_j}{\partial x_i}\right)\dfrac{\partial u_j}{\partial x_i}\end{bmatrix} \quad (23)$$

where  $\rho$ = fluid density (units unknown)
$t$ = time (units unknown)
$x_i$ = $i$th orthogonal axis of the coordinate space (units unknown)
$u_i$ = velocity components in the $x_i$ direction (units unknown)
$p$ = pressure parameter (units unknown)
$\mu$ = dynamic viscosity coefficient (units unknown)
$\lambda$ = second viscosity coefficient (units unknown)
$f_j$ = component of body force in the $x_j$ direction (units unknown)
$e$ = internal energy of the fluid (units unknown)
$k$ = thermal conductivity of the fluid (units unknown)
$T$ = temperature (units unknown)

The computerized process of CFD replaces the partial derivatives shown in Equations 21–23 with simpler discretized algebraic equations that are used to solve fluid flow values at discrete points in the system [Wala et al. 1997].

Several CFD simulations were conducted based on actual underground openings of a continuous mining operation. The computer program used by Wala et al. to execute the CFD for the airflow studies was CFD2000 version 2.2c. These studies produced airflow patterns that show the individual velocity vectors of airflow in underground mine openings and ventilation shafts. These patterns were experimentally validated using pressure and velocity measurements [Wala et al. 1997] and a new measurement method called particle image velocimetry (PIV) [Wala et al. 2001]. PIV uses a digital camera and a light source (usually a laser) to capture images of moving particles released into the airflow at specified intervals. At least two images are required to calculate the individual discrete flow vectors of the total airflow. This measurement method is described by Wala et al. [2001] and Turner et al. [2002]. PIV has been evaluated with CFD, but it has not yet been evaluated with experimental pressure and velocity measurements in mine openings [Turner et al. 2002].

Wala et al. only indicate the use of CFD at mine sites for airflow evaluation. However, they mention that CFD could be used for gas (i.e., methane) and dust dispersion, and they list several references pertaining to studies completed at mine sites that evaluated CFD for this possibility [Wala et al. 1997]. Additional studies have been done by Bennett et al. [2003a,b], which show that CFD has the ability to predict air contaminant concentrations in indoor occupational environments. Although not directly related to mining operations, these studies confirm that CFD can be used for gas and dust dispersion in underground mine openings.

### Brunner et al. Study

Brunner et al. [1995] completed a computer simulation to evaluate the use of CFD to calculate the airflow velocity required to prevent the backlayering of smoke from a mine fire against the ventilation flow [Brunner et al. 1995]. The computer simulation was completed on a vehicle mine fire wherein the vehicle was 2.4 m wide by 2.7 m high. The mine opening was 4.0 m wide by 4.5 m high by 200 m long. Three ventilation airflow velocity scenarios were executed: 0.5 m/sec, 0.75 m/sec, and 1.0 m/sec. Backlayering was predominant to 100 m against the ventilation airflow at an airflow velocity of 0.5 m/sec. At 0.75 m/sec airflow velocity, the backlayering was further diminished, and at 1.0 m/sec airflow velocity the backlayering was entirely eliminated [Brunner et al. 1995]. These results were not validated with field studies.

### Heerden and Sullivan Study

Heerden and Sullivan [1993] used CFD to evaluate dust suppression of continuous miners. This study was a computer simulation that calculated the airflow patterns in an underground entry containing a continuous miner. Individual airflow velocity vectors were plotted of the total airflow throughout the entry. Dust dispersion was determined qualitatively by assuming the dust particles would follow the individual airflow velocity vector, i.e., only trends in dust dispersion could be considered [Heerden and Sullivan 1993]. This simulation was used for different machine operational parameters and also for determining the effect of the rotating head of the miner on dust dispersion. The individual airflow vector results were stated to have been verified experimentally, although details of the comparison were not given. Applications for methane concentration dispersion were also discussed [Heerden and Sullivan 1993].

### Srinivasa et al. Study

Srinivasa et al. [1993] also completed a computer simulation for modeling airflow velocities and dust dispersion in an underground opening. This simulation was conducted on a longwall face and included the effects of the shearer and supports on the airflow. Dust particles were assumed to be inertialess and had no effect on the airflow patterns. Individual airflow velocity vector patterns were calculated using FIDAP, a CFD program created by Fluid Dynamics International (now available from Fluent, Inc.). The airflow patterns were calculated first, then the dust concentrations were calculated separately using the following equations. Particle motion was calculated as follows:

$$\rho^P\frac{\partial u_i^P}{\partial t}=F_D\left(u_i^F-u_i^P\right)+\left(\rho^P-\rho^F\right)g_i+\left(\rho^P-\rho^F\right)f_i \quad (24)$$

where subscripts/superscripts F and P refer to fluid phase and particle phase, respectively
$\rho$ = density (units unknown)
$t$ = time (units unknown)
$u_i$ = velocity in the $i$ direction (units unknown)
$f_i$ = body force vector (units unknown)
$g_i$ = gravitational force vector (units unknown)

$$F_D = 18 \frac{\mu^F}{D_P^2} C_D \qquad (25)$$

$\mu$ = dynamic viscosity (units unknown)
$C_D$ = drag coefficient (units unknown)
$D$ = particle diameter (units unknown)

The trajectory equation was given as:

$$\frac{\partial x_i}{\partial t} = u_i^P \qquad (26)$$

The advection-diffusion equation for the dispersed particles or pollutants was given as [Srinivasa et al. 1993]:

$$\rho \left[ \frac{\partial c}{\partial t} + u_i c_j \right] = \left( \rho \alpha c_j \right)_j + Q \qquad (27)$$

where $\rho$ = density (assumed to be particle density) (units unknown)
$t$ = time (units unknown)
$c_j$ = concentration of $j$th pollutant (units unknown)
$\alpha$ = mass diffusivity (units unknown)
$Q$ = a source term (units unknown)

Simulations were conducted on a longwall face, and the outcomes were compared with experimental results. The airflow velocities and the dust concentrations compared well across the cross-section of the entry perpendicular to the longwall face [Srinivasa et al. 1993]. Several simulations were conducted to test the effectiveness of an air curtain, semi-see-through curtains, and air-powered scrubbers in the walkway.

The simulation conducted on the air curtain demonstrated only a 15%–20% reduction of dust concentrations from the high concentration area to the walkway of the longwall section only if an air curtain quantity >0.05 m³/sec was maintained. It was stated that an airflow quantity >0.05 m³/sec for an air curtain is not practical [Srinivasa et al. 1993]. The air curtain simulation results were stated to correspond with the results of a study conducted by Hewitt in 1990 [Srinivasa et al. 1993; Hewitt 1990].

The simulation of the semi-see-through curtain maintained along the spill plate of the armored face conveyor demonstrated a 25%–30% reduction in dust concentrations from the face to the walkway. It was not stated if these results were verified experimentally [Srinivasa et al. 1993].

The simulation of the air-powered scrubber established that there was a 40%–50% protection efficiency associated with the scrubber at a distance of 2.1 m downstream [Srinivasa et al. 1993]. The simulation agreed to within 10% of the experimental results [Srinivasa et al. 1993]. No details of the experimental procedure were given.

As a result of this study, Srinivasa et al. concluded that CFD as a tool for predicting dust dispersion can be used to modify and improve dust control techniques.

## SURFACE MINE MODELS

The dust dispersion models used in surface mining have generally been adapted from existing industrial air pollution models. The surface models do not focus on a particular size fraction and are applicable to all particle sizes. All of these models, described below, are based on mathematical algorithms described earlier.

### MINE-SPECIFIC NUMERICAL MODELS

#### Cole and Fabrick Study

A number of models have been created for surface mining operations. Cole and Fabrick [1984] discuss pit retention of dust from surface mining operations. In their report, they mention a study completed by Shearer stating that approximately one-third of the emissions from mining activities escape the open pit. This is a very simplistic model that is representative of the box model algorithm. Further discussions are provided for a Gaussian plume model described by Winges. This model calculates the mass fraction of dust that escapes an open pit using the following mathematical model [Cole and Fabrick 1984]:

$$\varepsilon = \frac{1}{1 + \left( \frac{V_d}{K_z} \right) H} \qquad (28)$$

where $\varepsilon$ = mass fraction of dust that escapes an open pit
$V_d$ = particle deposition velocity (m/sec)
$K_z$ = vertical diffusivity (m²/sec)
$H$ = pit depth (m)

Fabrick also created an open-pit retention model based upon wind velocity at the top of the pit. This model, based on a Gaussian algorithm, is given as [Cole and Fabrick 1984]:

$$\varepsilon = 1 - V_d \left[ \frac{C}{u} \left( \frac{1}{2} + \ln \frac{w}{4} \right) \right] \quad (29)$$

where  $\varepsilon$ = mass fraction of dust that escapes an open pit
$V_d$ = particle deposition velocity (m/sec)
$u$ = wind velocity at the top of the pit (m/sec)
$C$ = empirical dimensionless constant equal to 7
$w$ = pit width (m)

The deposition velocity in both models was based on a gravitational settling velocity determined by Stoke's law. A comparison was completed using both models, and the results agreed well with each other and with the study by Shearer stating that one-third of the emissions from mining activities escape the open pit.

**TRC Environmental Study**

Several open-pit dust models are discussed in a study done for EPA by TRC Environmental Consultants [1995]. These include the models previously discussed by Cole and Fabrick. Another model, created by Herwehe in 1984, is a computer simulation using finite-element analysis. It takes into account many factors such as wind conditions, surface roughness, complex terrain, atmospheric stability, pollutant sources, particulate terminal settling and deposition velocities, and surface particulate accumulation [TRC Environmental Consultants 1995]. However, it was stated that this model may not give good results for open pits with pit angles greater than 35° from the horizontal or in stable atmospheres. This model also has not been tested with field results [TRC Environmental Consultants 1995]. Another model, the FEM (three-dimensional Galerkin finite-element model), which was not created specifically for the mining industry, was mentioned as one that could be modified for use in predicting dispersion of dust from open pits. Its drawback was that it required a very large computer to run the model. This model has also not been tested with field data. Both the Herwehe and FEM models use the concept of CFD in that they solve the Navier-Stokes equations for fluid motion to obtain particle dispersion results.

**Modeling of Blasting Phase Studies**

Modeling of dust dispersion for specific mining operations has been completed for the blasting phase through two studies. At the Kalgoorlie Consolidated Gold Mines Pty. Ltd., a computer program was created to determine dust dispersion from blasting operations. This program uses meteorology, bench height, blast design information, and rock density to predict the behavior of dust from blasting. It accounts for absorption of the dust on the pitwalls and for reflection of the dust off the pitwalls. The dust concentrations are calculated using settling velocities for different particle sizes and densities. The program is designed to determine if blasting will have an impact on a nearby town [Wei et al. 1999].

Another model for predicting dust dispersion from blasting operations has been created by Kumar and Bhandari [2002]. This model uses a gradient transport theory or an Eulerian approach, considering atmospheric stability and wind velocity and direction for computing dust concentrations at different distances from the blast [Kumar and Bhandari 2002]. Two integral equations are presented representing the dust dispersion model for the blast. However, none of the equation terms were defined in the study. No mention of any field validation of these models was given by Wei et al. or by Kumar and Bhandari.

**Pereira et al. Study**

Pereira et al. [1997] used a Gaussian dispersion equation to predict dust concentrations from the stockpiles of an operating surface mine in Portugal. The following equation is presented:

$$c = \frac{Q}{2\pi \sigma_y \sigma_z \bar{u}} \exp\left[-\frac{1}{2}\left(\frac{y_r}{\sigma_y}\right)^2\right] \exp\left[-\frac{1}{2}\left(\frac{h_e - z_r}{\sigma_z}\right)^2\right] \quad (30)$$

where
$c$ = pollutant concentration at location receptor = $(x_r, y_r, z_r)$ due to the emissions at source = $(0, 0, h_e)$ (units unknown)
$Q$ = emissions rate (units unknown)
$\sigma_y, \sigma_z$ = horizontal and vertical standard deviations or dispersion coefficients, respectively (units unknown)
$\bar{u}$ = average horizontal wind speed (units unknown)
$h_e$ = effective emissions height (units unknown)

This equation was used to create risk maps of air quality for locations surrounding the mine site. No experimental validation was performed to determine the accuracy of these maps to actual conditions [Pereira et al. 1997].

**ISC3 MODEL**

Dust dispersion modeling for surface mining operations, as required for air quality protection, is generally completed using an established model—the Industrial Source Complex model (ISC3) created by EPA. No other dust dispersion model has impacted the surface mining industry as much as the ISC3 model. This model was created by EPA to predict pollutant dispersion from industrial facilities and is available as a computer program from the EPA website [EPA 2005]. The pollutants for which it is designed include CO, $NO_x$, $SO_x$, VOC, Pb, and $PM_{10}$.

This model also includes a subroutine for modeling flat/complex terrain and has the ability to model dispersion from four types of emissions sources: point, which are typically stacks; volume, which are typically buildings; area, which are typically haul roads or storage piles; and open pit. In addition, the ISC3 model can calculate the deposition rates of $PM_{10}$ by using the deposition routine included in the model.

Modeling of dust dispersion at surface mining operations generally focuses on $PM_{10}$. The ISC3 model is particularly important to mining operations because state environmental permitting agencies often require its use for modeling dust dispersion from mining operations. The reason is that the ISC3 model is listed in 40 CFR 51, appendix W, as a "preferred air quality model." This designation means that this model can be used without a formal demonstration of applicability (40 CFR 51, appendix A of appendix W). An alternative model could possibly be used, but an extensive amount of justification would be required if an agency would even consider it for use. Therefore, a special review of the ISC3 model is presented here.

### ISC3 Modeling Technique

The ISC3 model is based on the Gaussian equation for point-source emissions, which is given as the following for the ISC3 model [EPA 1995c]:

$$\chi = \frac{QKVD}{2\pi u_s \sigma_y \sigma_z} \exp\left[-0.5\left(\frac{y}{\sigma_y}\right)^2\right] \quad (31)$$

where  $Q$ = pollutant emission rate (g/sec)
  $K$ = scaling coefficient to convert calculated concentrations to desired units (default value of $1\times10^6$)
  $V$ = vertical term (dimensionless)
  $D$ = decay term (dimensionless)
  $u_s$ = mean wind speed at release height (m/sec)
  $\sigma_y, \sigma_z$ = standard deviation of lateral and vertical concentration distribution (m)
  $\chi$ = hourly concentration at downwind distance $x$ ($\mu g/m^3$)
  $y$ = crosswind distance from source to receptor (m)

A series of supporting equations must be used with this equation. These equations are listed in appendix A of this report.

In order to describe how the ISC3 model works, a brief description of the modeling methodology is presented here. The ISC3 model calculates the $PM_{10}$ concentrations for receptor locations, which are based on a Cartesian coordinate system where each source and receptor has an X and a Y coordinate. These coordinates are input into the downwind and crosswind distance equations shown in Equations A–2 and A–3, respectively (see appendix A). These equations calculate the downwind distance $x$ and crosswind distance $y$, which are input into Equation 31.

Hourly meteorological data can be obtained from the EPA website [EPA 2005]. The data are acquired from meteorological monitoring stations operated by the National Weather Service or other individual facilities. The data are processed through another program called RAMMET, which is distributed by EPA along with the ISC3 model. This program uses the meteorological data to calculate the mixing height $z_i$. The mixing height is used in Equations A–9 or A–10. The program then organizes the data into a format readable by the ISC3 program. The data are then read into the ISC3 model for use in Equation 31.

Once all the data are entered, Equation 31 calculates the $PM_{10}$ concentration $\chi$ at the coordinates of the receptor. Generally, there is more than one receptor, and they are aligned in a grid format. $PM_{10}$ concentrations $\chi$ are calculated for each receptor point in the grid, with the emission source being stationary. These calculations are completed for each hour of available meteorological data. The modeled concentrations can be presented as hourly concentration values or as longer-term average concentrations, such as 24-hr averages or annual averages for a given receptor. The resulting grid of $PM_{10}$ concentrations allows for the creation of contour maps of the dispersion modeling results, where the contours represent the concentrations of $PM_{10}$. The user's guide by EPA [1995b,c] explains in greater detail how to operate the program.

It is important to characterize the emissions source correctly before conducting the dispersion modeling calculations. It must be determined if the source is a point, volume, area, or open-pit source, because mischaracterization of a source can impact modeled concentrations by an order of magnitude. In addition, the correct dimensions of the source must be input, as incorrect dimensions can cause large variances in modeled concentrations. The mischaracterization of the emissions source is a common cause of inaccuracy that can be eliminated if the source is characterized correctly. To eliminate this as a cause of inaccuracy, the National Stone, Sand, and Gravel Association (NSSGA) has begun work on a guidance document on characterizing sources [Heinerikson 2004].

### EPA Study

There have been very few studies completed to determine the ability of the ISC3 model to accurately predict $PM_{10}$ dispersion from surface mining operations. EPA completed a large-scale study at a surface coal mine in Wyoming in 1994–1995. This study, done in three phases [EPA 1994a,b; 1995a], reviewed the entire mining operation for dust dispersion. The emissions factors from the EPA's AP–42 were used to determine the amount of emissions from the operation. These emissions were then input into the ISC3 model to complete dispersion modeling. Field testing to validate the ISC3 model was completed by placing six $PM_{10}$ sampling stations throughout the surface mining operation. The sampling equipment used at each station was the Wedding $PM_{10}$ Reference Sampler. The six stations were used in addition to three existing $PM_{10}$ sampling stations located at the mine site to fulfill air quality permitting requirements [EPA 1994a]. The sampling stations were placed on both the upwind and downwind side of major excavating operations. Weather data were recorded throughout the duration of the test, and time studies of equipment operation were completed. The testing occurred over a period of 2 months, with air sampling occurring every other day [EPA 1994a]. The modeling results of the operations were compared to the actual measurements from the sampling network.

The EPA study documents that there is a significant overprediction of $PM_{10}$ emissions from the surface coal mining operation by the ISC3 model [EPA 1995a]. This report has a statistical protocol that defines significant overprediction as an overprediction of more than a factor of 2 at a single site where modeled versus measured results are compared [EPA 1994b]. No attempt to determine the source of the overprediction of $PM_{10}$ was made in this study. Consequently, it is not known whether the overprediction was caused by the emission estimation methods (AP–42) or by the dispersion model.

### Cole and Zapert Study

Cole and Zapert [1995] completed a study, submitted to the NSSGA, to test the ISC3 model at three Georgia stone quarries. It was stated that the ISC3 model had a history of overpredicting particulate concentrations based on data obtained by the U.S. Department of Energy's Hanford, WA, site [Cole and Zapert 1995]. This study calculated emission rates for operations, modeled the dispersion of the emitted particulates, and compared modeled versus measured particulate concentrations for each of the three stone quarries. The model testing methodology was similar to that used in the EPA study [EPA 1994a,b; 1995a]. The number and type of $PM_{10}$ sampling stations are unknown. However, it can be determined that there were at least two sampling stations at each site because there was a primary downwind site and a site located upwind of the prevailing winds to allow for subtraction of ambient $PM_{10}$ concentrations. Once the comparison of modeled versus measured results was completed, it was determined that the model overpredicted the actual $PM_{10}$ concentrations in a range of a factor of less than 1 (87% overprediction) to a factor of 5 [Cole and Zapert 1995].

The Cole and Zapert study concluded that there could be two reasons for the overprediction. One was that the ISC3 model failed at that time to account for any deposition of the particulates. The other reason was that the emissions factor for unpaved roads overpredicts the amount of emissions from haul trucks. The emissions factor was cited as the primary possible cause of overprediction because during the study it was noted that the hauling operations contributed 79%–96% of the $PM_{10}$ emissions from the entire quarrying operation [Cole and Zapert 1995].

EPA has been modifying a deposition routine for the ISC3 model, but no literature has been found where testing has been completed using the deposition routine in the ISC3 model. Cole and Zapert used an initial deposition routine created by EPA and found that it reduced the modeled results by 5%. However, even with this reduction in modeled $PM_{10}$ concentrations, there is still a significant overprediction. This has led the NSSGA to embark on a series of studies published during 1991–2001 that attempt to better quantify the $PM_{10}$ emissions from haul trucks [Richards and Brozell 2001].

### Reed et al. Studies

Reed et al. [2001] completed a study on the ISC3 model using a theoretical rock quarry. The study also concluded that hauling operations contributed the majority of $PM_{10}$ concentrations and that the haul truck emissions factors may be part of the cause of the overprediction of $PM_{10}$ concentrations by the ISC3 model. However, further analysis of the data provided by the Cole and Zapert study presented another hypothesis explaining the cause of the ISC3's overprediction. This hypothesis stated that since the majority of the sources producing $PM_{10}$ at surface mining operations are moving or mobile sources, the ISC3 model cannot accurately predict dust concentrations from mining operations because it is a model designed for predicting dust dispersion from stationary sources. This led to further investigations on dust dispersion modeling at surface mining operations, focusing on modeling the dispersion of dust generated from haul trucks.

Reed [2003] described a model called the Dynamic Component Program that can be used for predicting dust dispersion from haul trucks. The model is a modification of the ISC3 model created in Microsoft's Visual Basic and validated through testing at two surface mining operations. The model is based on a Gaussian equation similar to that used by the ISC3 model [Reed et al. 2002]:

$$\chi = \frac{QK}{2\pi w_s \sigma_y \sigma_z} \exp\left[-0.5\left(\frac{y}{\sigma_y}\right)^2\right] \qquad (32)$$

where  $Q$ = pollutant emission rate (g/sec)
$K$ = scaling coefficient to convert calculated concentrations to desired units (default value of $1\times10^6$)
$w_s$ = mean wind speed at release height (m/sec)
$\sigma_y, \sigma_z$ = standard deviation of lateral and vertical concentration distribution (m)
$\chi$ = hourly concentration at downwind distance $x$ ($\mu g/m^3$)
$y$ = crosswind distance from source to receptor (m)

Equation 32 is similar to Equation 31 used in the ISC3 model except that Equation 32 eliminates the use of the vertical and the decay terms. The major difference between the Dynamic Component Program and the ISC3 model is the methodology of applying the source emissions when predicting dust dispersion from that source.

The overprediction of dust dispersion by the ISC3 model occurs because the model applies the total emissions of the mobile sources to a specific area source. This application creates a constant uniform distribution of emissions over this specified area, as shown in Figure 1. In actual field conditions, however, the emissions from traffic or a mobile source are not uniform [Micallef and Colls 1999]. Figure 2 shows how the emissions from a moving haul truck actually occur. They act more like a moving point source rather than the continuous uniform emission distribution that the ISC3 model uses. A moving point source is more representative because the emissions occur abruptly as the emissions source approaches a point, then the emissions slowly dissipate as the source moves away from the point [Reed et al. 2002]. At a mining facility, this moving point source will move along a predictable path from the pit to the processing operations. The Dynamic Component Program uses the moving point-source methodology for predicting dust dispersion of haul trucks. This methodology should improve the ability to predict dust dispersion from haul trucks at surface mine sites.

Field studies at two surface mine locations—one stone quarry and one coal mine—were conducted to validate the results of the Dynamic Component Program with actual measurements. The results of the field studies showed that the Dynamic Component Program was an 85% improvement over the ISC3 model in predicting dust dispersion from haul trucks [Reed 2003]. However, the Dynamic Component Program is still in its infancy and can currently only predict dust dispersion of haul trucks traveling in a straight line. More work is required to perform modeling on haul trucks traveling haul roads that contain curves and to continue validating the model to ensure its accuracy.

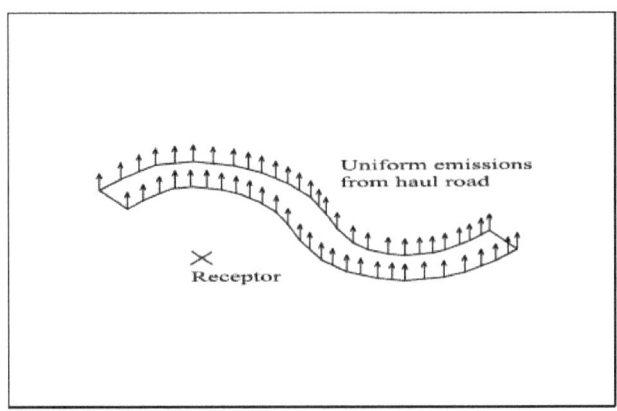

Figure 1.—Emissions from haul road as handled by the ISC3 model.

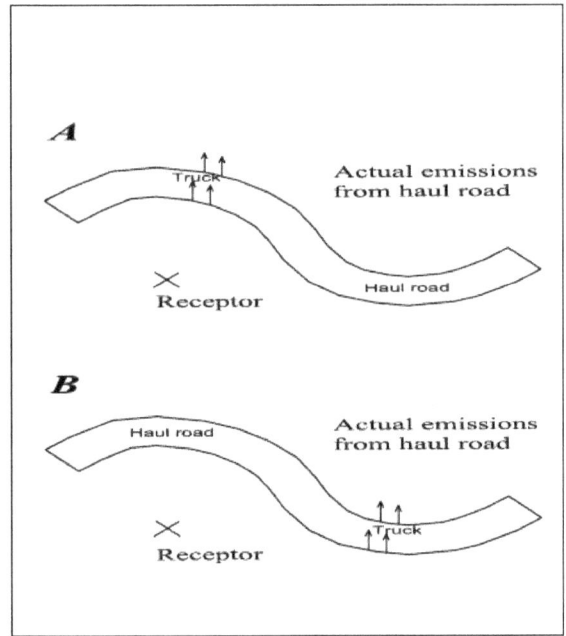

Figure 2.—Representation of actual emissions from a moving haul truck. A, arrows represent emissions at time $t_n$; B, arrows represent emissions at time $t_{n+1}$.

## ADDITIONAL MODELS

Two other dispersion models, although not developed specifically for the mining industry, have been used to model pollutant dispersion at surface mines: CALPUFF and AERMOD. These models are also available from the EPA website [EPA 2005].

CALPUFF is also an EPA-preferred air quality model. It is a Lagrangian model that uses continuous puffs to simulate emissions from sources [EPA 1998b]. It has applicability for many different pollutants, including $PM_{10}$. The ability of the CALPUFF model to simulate the effects of temporally and spatially varying meteorological conditions that occur more often over long pollutant transport distances makes it suitable for the prediction of long-range pollutant dispersion (>50 km). Conversely, the ISC3 model is appropriate for short-range pollutant dispersion because it requires constant steady-state meteorological conditions (<50 km) (40 CFR 51, appendix A of appendix W) [EPA 1998a]. The sources of data input into the CALPUFF model are similar to those of the ISC3 model, and the calculated results are concentrations that are output in a format similar to the ISC3 model (40 CFR 51, appendix A of appendix W). Studies have been conducted to verify the CALPUFF model results, but none have been conducted at mine sites. However, it has been determined that CALPUFF can produce results similar to the ISC3 model. Generally,

CALPUFF produces higher dispersion concentrations than those simulated by the ISC3 model for tall sources at close proximity to the source. However, at distances farther away from the source, the trend reverses, with the ISC3 model yielding higher dispersion concentrations than those simulated by CALPUFF [EPA 1998b].

AERMOD is the proposed replacement model for the ISC3 model. AERMOD's inputs and outputs are similar, and it has been improved significantly over the ISC3 model. Evaluations were conducted comparing AERMOD with the ISC3 model. These evaluations showed that, overall, AERMOD predicted pollutant concentrations that were closer to the actual measured concentrations than the ISC3 model. The two models produce relatively similar results for flat terrain. However, the largest amount of improvement was shown when comparing the two models in complex terrain. For the analysis conducted, AERMOD's short-term results varied from 0.67 to 1.12 times the observed results, whereas the ISC3 model's short-term results varied by 0.67 to 8.5 times the observed results [EPA 2003a]. Although AERMOD is a significant improvement over the ISC3 model, it is still in the beta-testing stage and has not yet been listed as a "preferred air quality model" in 40 CFR 51, appendix A of appendix W [EPA 2003b]. However, it is being distributed as AERMOD beta test version 02222.

## SUMMARY

A number of studies have been discussed in this report on mathematical modeling techniques, dust propagation models for underground mining, and dust propagation models for surface mining. Table 1 summarizes the dust dispersion models discussed in this report. A review of the dust dispersion models for surface mining shows that 12 different models have been created. Of these, only three have been tested at actual mining operations. The tested models are the Shearer model, which states that one-third of the mining emissions escape the open pit; the ISC3 model; and the Dynamic Component Program. The ISC3 model is the only model accepted by EPA for conducting dispersion modeling on surface mining facilities and has a documented history of use in the mining industry.

In contrast to the surface mining models, eight out of the nine models developed for underground mining operations have been field tested. It is notable that these models have been developed for underground coal operations except CFD, which has been adapted for underground mining use. It should also be noted that, unlike surface mining, there is no single accepted model for predicting dust dispersion in underground mining.

Table 1.—Summary of dust dispersion models

| Model | Algorithm type | Tested at mine site? | EPA-approved? |
|---|---|---|---|
| Underground: | | | |
| Hwang, Singer, and Hartz | Eulerian/Gaussian | No | No |
| Dawes and Slack | Gaussian | Yes | No |
| Courtney, Kost, and Colinet | Gaussian | Yes | No |
| Courtney, Cheng, and Divers | Eulerian | Yes | No |
| Courtney, Cheng, and Divers modified | Eulerian/Gaussian | Yes | No |
| Bhaskar | Eulerian | Yes | No |
| Chiang and Peng | Gaussian | Yes | No |
| Wala computational fluid dynamics (CFD) | Eulerian | Yes | No |
| Srinivasa | Eulerian | Yes | No |
| Surface: | | | |
| Shearer | Box | Yes | No |
| Winges | Gaussian | No | No |
| Fabrick | Gaussian | No | No |
| Herwehe finite-element | Eulerian | No | No |
| 3-D Galerkin finite-element model (FEM) | Eulerian | No | No |
| Kalgoorlie Consolidated Mines model | Unknown | Unknown | No |
| Kumar and Bhandari | Eulerian | No | No |
| Pereira, Soares, and Branquinho | Gaussian | No | No |
| ISC3 model | Gaussian | Yes | Yes |
| Dynamic Component Program [Reed 2003] | Gaussian | Yes | No |
| CALPUFF | Lagrangian | Tested, but not at mine site | Yes |
| AERMOD | Gaussian | Tested, but not at mine site | Pending |

The research completed for both the surface and underground models is important to the development of dust dispersion models. The evaluations conducted on underground mining models can be particularly important as they are a good basis for characterizing the prediction of dust concentrations and the deposition of dust. Underground mine openings are a more constant environment in that the airflow velocities and directions are less variable than on the surface. This "stable" (i.e., less varying) environment facilitates the characterization of the dust dispersion properties, whereas the evaluations conducted on surface mine operations can be more difficult to interpret due to the more variable environment. In either case, this important information can be applied to the future development of both underground and surface mining models. It is undoubtedly certain that future work will continue, as there is an ever-present need to improve upon the dust dispersion modeling of mining facilities.

# REFERENCES

Anderson JD Jr. [1995]. Computational fluid dynamics: the basics with applications. New York: McGraw-Hill.

Bennett JS, Crouch KG, Shulman SA [2003a]. Control of wake-induced exposure using an interrupted oscillating jet. Am Ind Hyg Assoc J $64$(1): 24–29.

Bennett JS, Feigley CE, Khan J, Hosni MH [2003b]. Comparison of emission models with computational fluid dynamic simulation and a proposed improved model. Am Ind Hyg Assoc J $64$(6):739–754.

Beychok MR [1994]. Fundamentals of stack gas dispersion. 3rd ed. Milton R. Beychok: Newport Beach, CA.

Bhaskar R [1987]. Spatial and temporal behavior of dust in mines: theoretical and experimental studies [Dissertation]. University Park, PA: The Pennsylvania State University, Department of Mineral Engineering.

Bhaskar R, Ramani RV [1989]. Dust flows in mine airways: a comparison of experimental results and mathematical predictions. In: Frantz RL, Ramani RV, eds. Publications produced in the Generic Mineral Technology Center for Respirable Dust in the year 1988. Washington, DC: U.S. Department of the Interior, Bureau of Mines, Office of Mineral Institutes, pp. 31–39.

Brunner DJ, Miclea PC, McKinney D, Mathur S [1995]. Examples of the application of computational fluid dynamics simulation to mine and tunnel ventilation. In: Proceedings of the Seventh U.S. Mine Ventilation Symposium (Lexington, KY, June 5–7, 1995). Littleton, CO: Society for Mining, Metallurgy, and Exploration, Inc., pp. 479–484.

California Air Resources Board [2003]. Ambient air quality standards for particulate matter. [http://www.arb.ca.gov/research/aaqs/pm/pm htm]. Date accessed: May 2005.

Carroll JM [1987]. Simulation using personal computers. Englewood Cliffs, NJ: Prentice-Hall, Inc.

CFR. Code of Federal regulations. Washington, DC: U.S. Government Printing Office, Office of the Federal Register.

Chiang HS, Peng SS [1990]. Development of dust distribution models for working faces. Vol. 1. Distribution model of airborne dust in longwall faces. Morgantown, WV: West Virginia University, Department of Mining Engineering. U.S. Bureau of Mines (USBM) grant No. G1135142, USBM open file report (OFR) 20–94.

Cole CF, Fabrick AJ [1984]. Surface mine pit retention. J Air Pollut Control Assoc $34$(6):674–675.

Cole CF, Zapert JG [1995]. Air quality dispersion model validation at three stone quarries. Englewood, CO: TRC Environmental Corp. TRC project No. 14884 for the National Stone Association, Washington DC.

Collett RS, Oduyemi K [1997]. Air quality modelling: a technical review of mathematical approaches. Meteorol Applic $4$(3):235–246.

Courtney WG, Cheng L, Divers EF [1986]. Deposition of respirable coal dust in an airway. Pittsburgh, PA: U.S. Department of the Interior, Bureau of Mines, RI 9041. NTIS No. PB 87–139424.

Courtney WG, Kost J, Colinet JF [1982]. Dust deposition in coal mine airways. Pittsburgh, PA: U.S. Department of the Interior, Bureau of Mines, Technical Progress Report (TPR) 116. NTIS No. PB 82–194853.

EPA [1994a]. Modeling fugitive dust impacts from surface coal mining operations: phase I. Research Triangle Park, NC: U.S. Environmental Protection Agency, Office of Air Quality Planning and Standards, Technical Support Division, EPA publication No. EPA–454/R–94–024.

EPA [1994b]. Modeling fugitive dust impacts from surface coal mining operations: phase II – model evaluation protocol. Research Triangle Park, NC: U.S. Environmental Protection Agency, Office of Air Quality Planning and Standards, Technical Support Division, EPA publication No. EPA–454/R–94–025.

EPA [1995a]. Modeling fugitive dust impacts from surface coal mining operations: phase III – evaluating model performance. Research Triangle Park, NC: U.S. Environmental Protection Agency, Office of Air Quality Planning and Standards, Emissions, Monitoring, and Analysis Division, EPA publication No. EPA–454/R–96–002.

EPA [1995b]. User's guide for the industrial source complex (ISC3) dispersion models. Vol. I. User instructions. Research Triangle Park, NC: U.S. Environmental Protection Agency, Office of Air Quality Planning and Standards, Emissions, Monitoring, and Analysis Division, EPA publication No. EPA–454/B–95–003a.

EPA [1995c]. User's guide for the industrial source complex (ISC3) dispersion models. Vol. II. Description of model algorithms. Research Triangle Park, NC: U.S. Environmental Protection Agency, Office of Air Quality Planning and Standards, Emissions, Monitoring, and Analysis Division, EPA publication No. EPA–454/B–95–003b.

EPA [1996]. Executive summary. In: Air quality criteria for particulate matter. Vol. I. Research Triangle Park, NC: U.S. Environmental Protection Agency, National Center for Environmental Assessment, EPA publication No. EPA/600/P–95/001aF, pp. 1–1 to 1–21.

EPA [1998a]. A comparison of CALPUFF modeling results to two tracer field experiments. Research Triangle Park, NC: U.S. Environmental Protection Agency, Office of Air Quality Planning and Standards, Emissions, Monitoring, and Analysis Division, EPA publication No. EPA–454/R–98–009.

EPA [1998b]. A comparison of CALPUFF with ISC3. Research Triangle Park, NC: U.S. Environmental Protection Agency, Office of Air Quality Planning and Standards, EPA publication No. EPA–454/R–98–020.

EPA [2003a]. AERMOD: Latest features and evaluation results. Research Triangle Park, NC: U.S. Environmental Protection Agency, Office of Air Quality Planning and Standards, Emissions, Monitoring, and Analysis Division, EPA publication No. EPA–454/R–03–003.

EPA [2003b]. Comparison of regulatory design concentrations: AERMOD vs ISCST3, CTDMPLUS, ISC-PRIME. Research Triangle Park, NC: U.S. Environmental Protection Agency, Office of Air Quality Planning and Standards, Emissions, Monitoring, and Analysis Division, EPA publication No. EPA–454/R–03–002.

EPA [2005]. Technology transfer network support center for regulatory air models. [http://www.epa.gov/scram001/tt22.htm]. Date accessed: May 2005.

Georgia Department of Natural Resources [1994]. Procedure to calculate a facility's "potential to emit" and to determine its classification. Atlanta, GA: Georgia Department of Natural Resources, Environmental Protection Division, Air Protection Branch, October 16.

Heerden J, Sullivan P [1993]. The application of CFD for evaluation of dust suppression and auxiliary ventilating systems used with continuous miners. In: Proceedings of the Sixth U.S. Mine Ventilation Symposium (Salt Lake City, UT, June 21–23, 1993). Littleton, CO: Society for Mining, Metallurgy, and Exploration, Inc., pp. 293–297.

Heinerikson A [2004]. Memorandum of June 3, 2004, from Arron Heinerikson, Trinity Consultants, Olathe, KS, to W. R. Reed, NIOSH Pittsburgh Research Laboratory, Pittsburgh, PA.

Hewitt A [1990]. Respirable dust sampling, research, and control in underground coal mines. In: Proceedings of Minesafe International Conference. Perth, Australia: Chamber of Mines and Energy of Western Australia.

Hwang CC, Singer JM, Hartz TN [1974]. Dispersion of dust in a channel by a turbulent gas stream. Pittsburgh, PA: U.S. Department of the Interior, Bureau of Mines, RI 7854. NTIS No. PB 233 492.

Kumar P, Bhandari S [2002]. Modeling of near source dust dispersal after surface mine blast in weak wind over undulated terrain in tropical conditions. In: Bandopadhyay S, ed. Application of Computers and Operations Research in the Mineral Industry: Proceedings of the 30th International Symposium. Littleton, CO: Society for Mining, Metallurgy, and Exploration, Inc., pp. 677–685.

Lippmann M [1995]. Size-selective health hazard sampling. In: Cohen BS, Hering SV, eds. Air sampling instruments for evaluation of atmospheric contaminants. 8th ed. Cincinnati, OH: American Conference of Governmental Industrial Hygienists, pp. 81–119.

Micallef A, Colls JJ [1999]. Measuring and modelling the airborne particulate matter mass concentration field in the street environment: model overview and evaluation. Sci Total Environ *235*(1–3):199–210.

NIOSH [1995]. NIOSH criteria for a recommended standard: Occupational exposure to respirable coal mine dust. Cincinnati, OH: U.S. Department of Health and Human Services, Public Health Service, Centers for Disease Control and Prevention, National Institute for Occupational Safety and Health, DHHS (NIOSH) Publication No. 95–106.

NIOSH [2002]. NIOSH hazard review: Health effects of occupational exposure to respirable crystalline silica. Cincinnati, OH: U.S. Department of Health and Human Services, Public Health Service, Centers for Disease Control and Prevention, National Institute for Occupational Safety and Health, DHHS (NIOSH) Publication No. 2002–129.

Oduyemi K [1994]. Modelling dispersion of pollutants with a simple software: it is a practical approach. In: Zanetti P, ed. Computer techniques in environmental studies V. Vol. 1. Pollution modeling. Southampton, U.K.: Wessex Institute of Technology Press, pp. 25–32.

Pereira MJ, Soares A, Branquinho C [1997]. Stochastic simulation of fugitive dust emissions. In: Baafi EY, Schofield NA, eds. Wollongong '96, Fifth International Geostatistics Congress. Vol. 2. Dordrecht, Netherlands: Kluwer Academic Publishers, pp. 1055–1065.

Ramani RV, Bhaskar R [1988]. Dust transport in mine airways. In: Frantz RL, Ramani RV, eds. Publications produced in the generic mineral technology center for respirable dust in the year 1984. Washington, DC: U.S. Department of the Interior, Bureau of Mines, Office of Mineral Institutes, pp. 25–31.

Reed WR [2003]. An improved model for prediction of $PM_{10}$ from surface mining operations [Dissertation]. Blacksburg, VA: Virginia Polytechnic Institute and State University, Department of Mining and Minerals Engineering.

Reed WR, Westman EC, Haycocks C [2001]. An improved model for estimating particulate emissions from surface mining operations in the eastern United States. In: Securing the Future – Proceedings of the International Conference on Mining and the Environment (Skellefteå, Sweden, June 25–July 1, 2001). Stockholm, Sweden: Swedish Mining Association, pp. 693–702.

Reed WR, Westman EC, Haycocks C [2002]. The introduction of a dynamic component to the ISC3 model in predicting dust emissions from surface mining operations. In: Bandopadhyay S, ed. Application of Computers and Operations Research in the Mineral Industry: Proceedings of the 30th International Symposium. Littleton, CO: Society for Mining, Metallurgy, and Exploration, Inc., pp. 659–667.

Richards J, Brozell T [2001]. Air quality monitoring and emission factor development: 1991–2001. Alexandria, VA: National Stone, Sand and Gravel Association.

Schnelle KB, Dey PR [2000]. Atmospheric dispersion modeling compliance guide. New York: McGraw-Hill.

Srinivasa RB, Baafi EY, Aziz NI, Singh RN [1993]. Three dimensional numerical modelling of air velocities and dust control techniques in a longwall face. In: Proceedings of the Sixth U.S. Mine Ventilation Symposium (Salt Lake City, UT, June 21–23, 1993). Littleton, CO: Society for Mining, Metallurgy, and Exploration, Inc., pp. 287–292.

TRC Environmental Consultants, Inc. [1995]. Dispersion of airborne particulates in surface coal mines: data analysis. Washington, DC: U.S. Environmental Protection Agency, Office of Air and Radiation, Office of Air Quality Planning and Standards.

Turner D, Wala AM, Jacob J [2002]. Experimental study of mine face ventilation system for validation of numerical models. In: De Souza E, ed. Proceedings of the North American/Ninth U.S. Mine Ventilation Symposium (Kingston, Ontario, Canada). Lisse, Netherlands: Balkema, pp. 183–189.

U.S. Department of Labor [1996]. Preventing silicosis. Washington, DC: U.S. Department of Labor, October 31.

Virginia Department of Environmental Quality [1996]. Business and industry guide to environmental permits in Virginia. Richmond, VA: Virginia Department of Environmental Quality.

Wala AM, Stoltz JR, Jacob JD [2001]. Numerical and experimental study of a mine face ventilation system for CFD code validation. In: Proceedings of the Seventh International Mine Ventilation Congress (Krakow, Poland, June 17–22, 2001), pp. 411–417.

Wala AM, Yingling JC, Zhang J, Ray R [1997]. Validation study of computational fluid dynamics as a tool for mine ventilation design. In: Ramani RV, ed. Proceedings of the Sixth International Mine Ventilation Congress (Pittsburgh, PA, May 17–22, 1997). Littleton, CO: Society for Mining, Metallurgy, and Exploration, Inc., pp. 519–525.

Watson JG, Chow JC, DuBois D, Green M, Frank N, Pitchford M [1997]. Guidance for network design and optimum site exposure for $PM_{2.5}$ and $PM_{10}$. Research Triangle Park, NC: U.S. Environmental Protection Agency, Office of Air Quality Planning and Standards.

Wei Y, Ayotte K, Howard R [1999]. KCGM blasting dust modelling and management system. In: CSIRO Exploration and Mining Research Review. Commonwealth Scientific and Industrial Research Organisation, Division of Exploration and Mining, Australia, pp. 204–205.

Xu L, Bhaskar R [1995]. A simple model for turbulent deposition of mine dust. In: Proceedings of the Seventh U.S. Mine Ventilation Symposium (Lexington, KY, June 5–7, 1995). Littleton, CO: Society for Mining, Metallurgy, and Exploration, Inc., pp. 337–343.

# APPENDIX A.—SUPPORTING EQUATIONS USED WITH THE ISC3 MODEL

The following equations are all required in order to use the Gaussian equation for pollutant dispersion. This information is taken directly from EPA [1995c].

**The Gaussian equation for point source emissions**

$$\chi = \frac{QKVD}{2\pi u_s \sigma_y \sigma_z} \exp\left[-0.5\left(\frac{y}{\sigma_y}\right)^2\right] \tag{A-1}$$

where  $Q$ = pollutant emission rate (g/sec)
$K$ = scaling coefficient to convert calculated concentrations to desired units (default value of $1\times10^6$)
$V$ = vertical term (dimensionless)
$D$ = decay term (dimensionless)
$u_s$ = mean wind speed at release height (m/sec)
$\sigma_y, \sigma_z$ = standard deviation of lateral and vertical concentration distribution (m)
$\chi$ = hourly concentration at downwind distance $x$ ($\mu g/m^3$)
$y$ = crosswind distance from source to receptor (m)

**Downwind and crosswind distances**

$$x = -(X(R) - X(S))\sin(WD) - (Y(R) - Y(S))\cos(WD) \tag{A-2}$$

where  $x$ = downwind distance (m)
$X(R)$ = x coordinate of receptor (m)
$Y(R)$ = y coordinate of receptor (m)
$X(S)$ = x coordinate of source (m)
$Y(S)$ = y coordinate of source (m)
$WD$ = direction from which wind is blowing (angle measured clockwise from north) (degrees)

$$y = -(X(R) - X(S))\cos(WD) - (Y(R) - Y(S))\sin(WD) \tag{A-3}$$

where  $y$ = crosswind distance (m)
$X(R)$ = x coordinate of receptor (m)
$Y(R)$ = y coordinate of receptor (m)
$X(S)$ = x coordinate of source (m)
$Y(S)$ = y coordinate of source (m)
$WD$ = direction from which wind is blowing (angle measured clockwise from north) (degrees)

## Wind speed profile

$$u_s = u_{ref}\left(\frac{h_s}{z_{ref}}\right)^p \tag{A-4}$$

where  $u_s$ = mean wind speed at release height (m/sec)
$u_{ref}$ = observed wind speed from a measured reference height ($z_{ref}$) (m/sec)
$h_s$ = stack height (m)
$p$ = wind profile exponent (dimensionless)
$z_{ref}$ = measured reference height for wind speed (m)

**Default values**

| Stability category | Rural exponent | Urban exponent |
|---|---|---|
| A | 0.07 | 0.15 |
| B | 0.07 | 0.15 |
| C | 0.10 | 0.20 |
| D | 0.15 | 0.25 |
| E | 0.35 | 0.30 |
| F | 0.55 | 0.30 |

## Plume rise due to momentum

$$h_e = h'_s + 3d_s\left(\frac{v_s}{u_s}\right) \tag{A-5}$$

where  $h_e$ = plume rise (m)
$h'_s$ = stack height (m)
$d_s$ = inside diameter of stack (m)
$v_s$ = exit velocity of stack gas (m/sec)
$u_s$ = mean wind speed (m/sec)

## Dispersion parameters ($\sigma_y$, $\sigma_x$)

$$\sigma_y = 465.11628(x)\tan(TH) \tag{A-6}$$

where  $TH = 0.017453293[c - d\ln(x)]$ \quad (A-7)

where  $x$ = downwind distance (km)
$c,d$ = coefficients

**Coefficient default values**

| Pasquill stability category | c | d |
|---|---|---|
| A | 24.167 | 2.5334 |
| B | 18.333 | 1.8096 |
| C | 12.500 | 1.0857 |
| D | 8.333 | 0.72382 |
| E | 6.250 | 0.054287 |
| F | 4.1667 | 0.36191 |

The Pasquill stability category refers to the stability of air layers near the ground. It is based upon wind speed and insolation (incoming solar radiation) [Schnelle and Dey 2000]. The following table defines the six categories.

**Pasquill-Gifford stability categories**

| Surface wind (measured at 33 ft) (mph) | Daytime insolation | | | Nighttime cloudiness | |
|---|---|---|---|---|---|
| | Strong | Moderate | Slight | Thinly overcast or ≥4/8 cloudiness | ≤3/8 cloudiness |
| 4.5 | A | A–B | B | — | — |
| 4.5–6.7 | A–B | B | C | E | F |
| 6.7–11.2 | B | B–C | C | D | E |
| 11.2–13.4 | C | C–D | D | D | D |
| >13.4 | C | D | D | D | D |

A  Extremely unstable.    B  Moderately unstable.    C  Slightly unstable.
D  neutral.              E  Slightly stable.        F  Moderately stable.

NOTE.—Insolation is the rate of radiation from the sun received per unit of earth's surface. Strong insolation corresponds to sunny midday in summer. Slight insolation corresponds to similar conditions in winter. For A–B, B–C, and C–D, average values are taken. Night refers to 1 hr before sunset to 1 hr after dawn. Regardless of wind speed, the neutral category D should be assumed for overcast conditions during day or night and for any sky conditions during the hour preceding or following night.
Source: Schnelle and Dey [2000].

$$\sigma_z = ax^b \tag{A-8}$$

where   $x$ = downwind distance (km)
        $a, b$ = coefficients

**Default values**

| Pasquill stability category | $x$ (km) | $a$ | $b$ | Pasquill stability category | $x$ (km) | $a$ | $b$ |
|---|---|---|---|---|---|---|---|
| A | <0.10 | 122.800 | 0.94470 | E | <0.10 | 24.260 | 0.83660 |
|   | 0.10–0.15 | 158.080 | 1.05420 |   | 0.10–0.30 | 23.331 | 0.81956 |
|   | 0.16–0.20 | 170.220 | 1.09320 |   | 0.31–1.00 | 21.628 | 0.75660 |
|   | 0.21–0.25 | 179.520 | 1.12620 |   | 1.01–2.00 | 21.628 | 0.63077 |
|   | 0.26–0.30 | 217.410 | 1.26440 |   | 2.01–4.00 | 22.534 | 0.57154 |
|   | 0.31–0.40 | 258.890 | 1.40940 |   | 4.01–10.00 | 24.703 | 0.50527 |
|   | 0.41–0.50 | 346.750 | 1.72830 |   | 10.01–20.00 | 26.970 | 0.46713 |
|   | 0.51–3.11 | 453.850 | 2.11660 |   | 20.01–40.00 | 35.420 | 0.37615 |
|   | >3.11 | — | — |   | >40.00 | 47.618 | 0.29592 |
| B | <0.20 | 90.673 | 0.93198 | F | <0.20 | 15.209 | 0.81558 |
|   | 0.21–0.40 | 98.483 | 0.98332 |   | 0.21–0.70 | 14.457 | 0.78407 |
|   | >0.40 | 109.300 | 1.09710 |   | 0.71–1.00 | 13.953 | 0.68465 |
| C | All | 61.141 | 0.91465 |   | 1.01–2.00 | 13.953 | 0.63227 |
| D | <0.30 | 34.459 | 0.86974 |   | 2.01–3.00 | 14.823 | 0.54503 |
|   | 0.31–1.00 | 32.093 | 0.81066 |   | 3.01–7.00 | 16.187 | 0.46490 |
|   | 1.01–3.00 | 32.093 | 0.64403 |   | 7.01–15.00 | 17.836 | 0.41507 |
|   | 3.01–10.00 | 33.504 | 0.60486 |   | 15.01–30.00 | 22.651 | 0.32681 |
|   | 10.01–30.00 | 36.650 | 0.56589 |   | 30.01–60.00 | 27.074 | 0.27436 |
|   | >30.00 | 44.053 | 0.51179 |   | >60.00 | 34.219 | 0.21716 |

Source: EPA [1995c].

## Vertical term

$$V = \exp\left[-0.5\left(\frac{z_r - h_e}{\sigma_z}\right)^2\right] + \exp\left[-0.5\left(\frac{z_r + h_e}{\sigma_z}\right)^2\right]$$
$$+ \sum_{i=1}^{\infty}\left\{\exp\left[-0.5\left(\frac{H_1}{\sigma_z}\right)^2\right] + \exp\left[-0.5\left(\frac{H_2}{\sigma_z}\right)^2\right] + \exp\left[-0.5\left(\frac{H_3}{\sigma_z}\right)^2\right] + \exp\left[-0.5\left(\frac{H_4}{\sigma_z}\right)^2\right]\right\} \quad \text{(A-9)}$$

where
- $h_e = h_s + \Delta h$
- $h_e$ = plume height (m)
- $h_s$ = stack height (m)
- $\Delta h$ = plume rise (m)
- $H_1 = z_r - (2iz_i - h_e)$
- $H_2 = z_r + (2iz_i - h_e)$
- $H_3 = z_r - (2iz_i + h_e)$
- $H_4 = z_r + (2iz_i + h_e)$
- $z_r$ = receptor height above ground (m)
- $z_i$ = mixing height (m)

$$V = \frac{\sqrt{2\pi}\sigma_z}{z_i} \quad \text{(A-10)}$$

This form (Equation A–10) is used to save on computational time without sacrificing accuracy.

There are variations of the vertical term depending on the routine used in calculating dispersion.

## Decay term

$$D = \exp\left(-\psi \frac{x}{u_s}\right) \quad \text{For } \psi > 0 \quad \text{(A-11)}$$

$$D = 1 \quad \text{For } \psi = 0$$

where $\psi$ = decay coefficient

$$\psi = \frac{0.693}{T_{1/2}} \quad \text{(A-12)}$$

- $T_{1/2}$ = pollutant half life (sec)
- $x$ = downwind distance (m)
- $u_s$ = mean wind speed at release height (m/sec)

Default value of $\psi = 0$ unless specified.

www.ingramcontent.com/pod-product-compliance
Lightning Source LLC
Chambersburg PA
CBHW081822170526
45167CB00008B/3506